Sheep Keeping

Richard Spencer

Sheep Keeping

*Inspiration and practical advice
for would-be smallholders*

 National Trust

First published in the United Kingdom in 2010 by
National Trust Books
10 Southcombe Street
London W14 0RA
An imprint of Anova Books Company Ltd

ISBN 9781905400874

A CIP catalogue for this book is available from the British Library.
15 14 13 12 11 10
10 9 8 7 6 5 4 3 2 1

Reproduction by Mission Productions Ltd, Hong Kong
Printed and bound by WS Bookwell OY, Finland

This book can be ordered direct from the publisher at the website
www.anovabooks.com, or try your local bookshop. Also available
at National Trust shops, including www.nationaltrustbooks.co.uk.

The information contained in this book is not a substitute for
specialist advice on specific issues concerning sheep keeping. Neither
the publishers, the National Trust nor the author make any warranties
or representations regarding the accuracy of material contained herein
and exclude liability to the fullest extent permitted by law for any
consequence resulting from reliance upon it. Reasonable care has been
taken to ensure that information relating to laws and regulations reflects
the general state of the law as understood by the author at the time of
going to press. Any liability for inaccuracies or errors contained herein is
expressly excluded to the fullest extent permitted by law. If you intend to
rear and keep sheep for any purpose you must contact Defra as well as
your local Animal Welfare Office and ensure that you comply with all
applicable legislation.

CONTENTS

SO, YOU WANT TO KEEP SHEEP?

Many of us are only two or three generations removed from a relative who cared for livestock, so keeping farm animals is not such a big step to take.

WHY CHOOSE SHEEP?

The very first fact you need to accept is that, as every shepherd knows, a sheep is the only animal on the earth looking for the quickest way to die; they do seem to be constantly getting into mischief of some sort. Having said that, I love sheep; in my experience they are the most fascinating of any animals – frustrating, annoying, perhaps, but the source of so much pleasure. If ever you are feeling lonely and unloved, feed some hungry sheep!

What is more natural than keeping livestock on your acres? There are a number of options available. What about goats? Well, I have kept them and have seen how they will eat anything – the hedge, your garden, next door's garden, your hair, the washing on the line, anything in your pockets – in fact, anything except grass. So goats are a risky option. Pigs are Mother Nature's answer to the plough and love nothing more than turfing up your land. Cows are big, expensive and in a wet winter your field would look like the local rugby pitch at the end of the season.

All that is left, then, is sheep … and why not? What is more British than a flock of sheep on a sunny afternoon lying in dappled shade under the hedge and the oak tree, chewing the cud? And it is infinitely more pleasant knowing that it is your field and that you put the sheep there. It really is as good as it sounds, but, before you commit to keeping sheep, you need to know that it is almost as big a responsibility as having children. You must always keep an eye on sheep and be aware of their needs; this can be a problem – when you wish to go away, for example – but it is not insurmountable.

WHAT TYPE OF LAND?

Whatever livestock you are keeping, the type of land you have is important, because it affects the way you manage your animals. As long as sheep can 'lie dry' they will do well. For sheep to be able to lie dry the land needs to be free-draining – on sandy soil or perhaps slightly raised ground. Although it is not free-draining, a clay soil can, to an extent, lie dry if it is slightly raised or is slightly under-grazed with tussocks of dead grass on a mature, thick turf.

If the area of land for grazing is limited, clay soil can very quickly become a quagmire in wet weather, particularly once the matt of the turf has been grazed away. Free-draining sandy soil can equally become poached (trodden up) in wet weather if stocking levels are too high and sheltered areas are not provided. As a rule of thumb, a commercial farmer, who will probably have sufficient areas of land around which to rotate his grazing, will keep five ewes per acre. On a more limited area, where the number of secure grazing acres is restricted, a lower stocking density is to be recommended – perhaps two to three Down or commercial ewes per acre, or three to four primitives, such as Hebridean, Soay or Manx Loghtan (for

7

more information on individual breeds, see pages 24–42).
Remember, all land is different. Start slowly and find the
level that suits you, your chosen sheep and your grassland.

On my land, we farm a heavy clay soil and find that sheep
thrive on it. The secret is to make sure that, when the spring
weather arrives and the ground begins to dry, the sheep have
been moved to other pastures. The poached (trodden up) land
will heal remarkably well. In severe cases, scatter a handful of
grass seed over the bare ground; it is surprising how quickly
the grass will grow and it provides competition for the weeds.

PROVIDING SHELTER

Shelter is important. Every point on any landscape is subject
to a prevailing wind. If there are no hedges to provide shelter
and perhaps the land could be more free-draining, during the
wetter and colder winter months the sheep will need to be
provided with housing or shelters, which they will find and
use if they need to, sheep being amazingly wise when it comes
to surviving inclement weather.

Regardless of the number of sheep, the type of housing or
shelter required depends on what time of year lambing occurs.
If lambing takes place in April or May the weather is usually
sufficiently mild for all sheep to lamb outside. However, there
are certain tasks that need to be completed as a matter of
routine with your flock. You should have a roofed area
sufficient to accommodate your sheep – it keeps them and
you dry should it be raining and provides shade in the hot
summer months. If you have a pedigree flock, a roofed area
keeps the paperwork, notebook or laptop dry.

In the long term, of course, you may wish to consider planting a hedge, but remember that it must be fenced in a stock-proof manner, because most sheep will browse unprotected hedges to extinction (for more on hedging, see page 13).

WHAT ARE YOUR PLANS FOR YOUR SHEEP?

Before making the final decision that you are going to keep sheep, serious thought must be given as to what you plan to do with them. What do you want them for? For example, if you have a small paddock that is overlooked by the house and garden, some of the more picturesque breeds would be ideal, such as coloured breeds, breeds with horns or Longwools with their 'dreadlock' fleeces (for more information on individual breeds, see pages 24–42). Of course, all sheep kept in flocks of the same breed look picturesque, so it is really down to what catches your eye.

In the spring when the sun is shining, the grass begins to grow and the browns of an over-wintered pasture are transformed to a lovely shade of green, the new leaves and blossoms provide a picture that is second to none. The view is complete when, every time you look out of the window, you see new-born lambs of your chosen breed lying in the sun, probably on mum's back or racing around the pasture.

The other scenario is a smallholding with a number of acres to keep under control. This slightly alters the requirement in terms of handling facilities. While three or four ewes and their lambs can be controlled with a few hurdles and caught individually – even the larger breeds with a little determination – a larger area of land, say 2–4ha (5–10 acres), could support between 20 and 40 ewes. A small number of sheep can be coaxed (fooled) into coming inside with a bucket of food and you can close the gate behind them once they are in. However, a larger flock will require a dog, which can be very expensive (around £500 to £1,500), or some very obliging family, neighbours and friends, whose tolerance may be sorely tested. If they have previous experience of chasing sheep for a couple of hours, it is amazing how suddenly they will remember a prior commitment when called upon a second time.

Assuming that your flock has been brought into the handling area, hurdles and physical effort can still suffice. Agricultural suppliers have a large range of 'draughting' gates, races (gated, sheep-width passageways) and handling systems available, which takes the physical effort out of routine work, but they need to be arranged in the correct place. Ideally the sheep need to run uphill towards light as they move through a race. Only experience, trial and error will tell you how best to arrange your system. In my own experience, a 'three-way' system is ideal – as the sheep pass down the race, they can be diverted into one of three directions as they pass through the gate. However, it is essential to be able to identify the animals quickly and have fast reactions.

With our own flock we find the best method is to bring the entire mob into a yard and then hold a small group of 10–20 in a tight pen. It is easy to move through them, marking each one as you dose it, or slip it through the draughting gate. Whenever holding flocks of sheep in an enclosed space, you must always make sure that small ewes, older ewes or lambs are not being crushed at the back or in a corner.

LEGAL REQUIREMENTS

There are literally millions of sheep in the UK.
To help control, isolate and eliminate disease in case
of an outbreak, it is essential to comply with certain
record-keeping requirements.

REGISTRATION AND CHIPPING

Keeping livestock necessitates compliance with the law.
Your land will need to be registered with Defra (Department
for Environment, Food and Rural Affairs), who will allocate
a holding number to you (see Useful Addresses, page 88).
All the animals must be tagged, and a record kept of tag
replacements should a tag be lost. As well as your holding
number, Defra will give you a flock designation number,
which will be 'UK', followed by a series of six numbers, and
then the number of your sheep. Suppliers of tags advertise
in all livestock-related journals and your local agricultural
merchant will be able to give you further advice.

At the time of writing, 'chip' tags are about to become
a legal requirement for animals over 12 months of age;
in other words those that have been selected for breeding.
The idea is that every time a sheep is moved the tag is
scanned by electronic scanner (about £350 to purchase), the
information transferred to a computer and then e-mailed to
the appropriate government office. However, manual reading
of the tags and transfer information will still be accepted by
post. All sheep keepers need plenty of sheep-movement forms
from their local Trading Standards Office, and these need to be
completed every time a sheep moves on or off the holding. A

copy must be sent to the Trading Standards office within three days of movement.

Every movement of sheep on or off the holding must also be recorded in your animal movement record, which needs to be kept for several years – Defra can advise as to the details. Anyone moving livestock must register with their local authority and obtain a certificate. They should also take a written exam to be able to legally transport the sheep. It is also a legal requirement to notify Defra if your sheep contract certain diseases (for more on this, see page 49).

MEDICINES RECORD

All medicinal and antibiotic treatments must be recorded in a medicines book together with animal ID, date of treatment, duration of treatment, meat-withdrawal period, details of the problem needing treatment, name of medicine, quantity of medicine, type of administration (intramuscular, subcutaneous or intravenous), response after seven days, batch number of medicine, date of purchase and date of expiry for every treatment.

FENCING YOUR LAND

Maintaining your fencing seems to be a never-ending task –
there always seems to be something that needs doing.
To get the best results, it is essential that you choose suitable
fencing to secure your land.

Once you have made the decision to keep sheep and have selected the breed you want to buy, before you go any further you need to make sure that the fencing around your perimeter is secure. Keeping sheep on poorly fenced ground is a sure-fire, guaranteed way to fall out with your neighbours. If your sheep escape – and they are renowned for their ability to squeeze through the smallest of holes – returning them to base can prove rather difficult. When you wish to move them from point A to point B, they can and often do refuse to pass through an open gateway. They quickly forget that they are supposed to eat grass, happily devouring anything within sight, from tulips and primrose to rosebuds. Even worse, they could be flattening a farmer's cereal crop or possibly wandering through a mature crop of maize a couple of metres tall, crunching on the cobs and proving very difficult to find. So, good fencing is essential … what are the options?

HEDGING

Hedges provide shelter in times of inclement weather, shade from the sun and the delight of planting standard trees in your hedge line – think of crab apples, horse chestnut, hawthorn, wild cherry, bullace, guelder rose, or perhaps even an oak tree. All are native trees, providing a huge variety of leaf colour and the most beautiful spring flowers.

A managed mature hedge is the best option for fencing sheep. However, if you have a hedge that has been neglected and is in effect a haphazard row of small trees, then it needs to be pleached (see Glossary, page 94), which will cost £10–£15 per metre and possibly more, depending on how overgrown it has become.

If you are planting a hedge, Defra can advise on which plants are most suitable. A combination of hawthorn as the main deterrent, with some dog rose, honeysuckle and field maple, is effective. Hazel is another option to include. Blackthorn is not recommended because the thorn can cause septicaemia.

FENCING

It is paramount that no matter how good the hedge, at the very least an electric fence with two or three strands of battery-powered electrified wire or netting should be erected inside the hedge. However, this is really only a temporary measure and you can guarantee that if a sheep becomes caught in the electric wire it will be at the point furthest

away from the off switch. Trying to untangle a sheep from the wire while the power is still switched on is not to be recommended. A better investment is a post-and-wire fence. Yes it will cost money, but it will last for years.

Regardless of the abundance of grass in your field, eventually sheep will manage to either put their heads through the stock netting and get stuck or will tangle their horns in the wire, presumably because of the attraction of the hedging. It is not uncommon (in other words, it has happened to me) to release a lamb from the wire, only to return five minutes later to find it stuck in the same place. To remove, turn the sheep upside down (this is not easy) and, in effect, ease the head out forwards. An old method of preventing stop sheep from entangling their horns in wire was to tie a plank of wood across the horns, but this is no longer recommended for animal welfare reasons.

Sheep will nibble on leaves on hedges to a greater or lesser degree. A fence will allow the hedge to develop and furthermore will stop sheep that have taken shelter from getting too close and rubbing themselves against the main stems; eventually the natural grease in the fleece is transferred to the hedge plants and will smother the plant, causing the hedge to become open at the bottom, restricting growth and effectively destroying the shelter and probably the hedge.

GATES

Once your hedge and fencing are up to the required standard, it is now time to pay attention to your gates. The best way to hang boundary gates is to have the top gate hook pointing downwards and the bottom gate hook pointing upwards. This is a little difficult to do but, equally, it is very difficult to remove the gate and consequently a deterrent for those who might have an eye on your sheep – or an eye on your new gate, which can cost as much as £70 or more.

POST-AND-RAIL FENCING

Another option, which is very expensive indeed, is post-and-rail fencing; in some circumstances, it may be your best choice. However, it will rarely blend in as well as a stock netting wire fence constructed to protect a hedge, because the hedge will tend to grow up and through the wire, the fence becoming virtually invisible in only a couple of years.

THE HISTORY
AND DEVELOPMENT OF
SHEEP BREEDING

*To follow the history of sheep is almost to follow the
social, cultural and economic development of Britain, and
is a must for anyone who wishes to understand the
characteristics of the different breeds.*

DIFFERENT BREEDS; DIFFERENT CHARACTERISTICS

There are over 60 pure breeds of sheep in the UK, with
countless popular commercial crossbreeds. The most popular
crossbreed is the Mule, which can take a number of forms,
but is always sired by a Blue-faced Leicester ram. Different
sheep breeds have obviously different characteristics and the
most important of these are the breeding characteristics. It is
self-evident that the whole point of keeping sheep is for the
lambs they eventually produce; you have kept your sheep for
12 months and this is their one chance to bring you some
financial return for all your efforts. Your best hope is to walk
out to the sheep for your early morning inspection to find a
ewe with loads of milk and two healthy lambs. Then, as you
pick the lambs up to bring them inside for 24 hours, you have
mum butting your knees from behind as she follows you and
her lambs inside. Mules are brilliant for this kind of mothering
characteristic. What you do not want, however, is to inspect
your sheep at 2am on the first wet night for weeks to find
a shearling ewe nervously inspecting something that has
mysteriously appeared on the ground behind her, jumping

backwards every time it twitches and as soon as you
pick up the lamb, racing into the remainder of the flock
at the far side of the field. Don't panic; there are good
and bad traits in all breeds and, equally, there are good
and bad mothers.

THE HISTORY OF SHEEP KEEPING

The history of sheep is quite fascinating, and many books
have been written on the subject. Agriculture in Britain has
been developing for centuries, with its progress accelerating
in the 19th century. Due to the lack of a communication
network in early times, breeds of livestock tended to be
particular to one area, which explains the diversity that
is still apparent today.

EARLY BREEDS

The Romans are believed to have had some influence on
sheep breeds, and the Longwools are thought to have been
introduced by them: classic examples are the Cotswold,
Lincoln and Leicester Longwools. The Vikings are thought to
have been responsible for the introduction of what are known
as the primitive breeds, such as the Hebridean, Soay, Manx
Loghtan and Shetland. As their names suggest, these breeds
became localised on remote islands and the absence of outside
genetic interference resulted in breeds becoming very uniform
in type and unchanged for centuries.

At some stage in the late Middle Ages, the Jacob was
introduced from Europe; this breed has remained pure and
there is little, if any, evidence to suggest it has been involved

in any crossbreeding or development, probably due to its chocolate brown and white markings. These markings are very striking to look at, but as early as the 14th century British wool was becoming popular in continental Europe for creating cloth and, since you can only dye white wool, the Jacob would have been of limited popularity. Its primary function, then, was to look attractive, grazing the parkland of the landed gentry, and indeed a mature four-horned Jacob ram does look absolutely magnificent.

It is also believed that a very important factor in relation to sheep breeding – namely 'out of season lambing' – came about due to an unfortunate set of circumstances. Such was the confidence of the Spanish when the Armada set sail in 1588 that they included sheep in the cargo on their ships. Were they 'on board' to provide a supply of breeding stock once the successful Armada landed and Britain was invaded? More probably they were a convenient supply of food. Whatever the case, when Sir Francis Drake had finished his game of bowls and turned his attention to the Spanish problem, some of the Armada's ships foundered on the south coast of England and the 'Portland' sheep swam ashore. They can breed at any time of year and this characteristic is to be found particularly in the Dorset Down, and very markedly in the Dorset Horn and Polled Dorset, these latter two only differing in the presence or absence of horns.

GEOGRAPHICAL INFLUENCE

To add to the mish-mash of breed origins, one must also factor in the geographical influence of the British landscape on breeds of sheep, which have developed to suit their environment and the farming systems that grew up around them.

Historically, hill country has small, hardy sheep, which are nimble on their feet to jump around the steep slopes and thrive by nibbling the sparse vegetation. Breeds include the Welsh Mountain (white) and Black Welsh Mountain, who have a harsh, tough fleece to withstand the winter weather and a low lambing percentage, at probably little more than 100 per cent. (Lambing percentage refers to the number of lambs born; 167 lambs born to 100 ewes is 167 per cent. Due to the harsh environment on the hills, a lambing percentage of around 100 per cent is acceptable. On the milder lowlands however, around 200 per cent would be ideal.)

Moorland sheep also have a harsh, thick fleece but perhaps slightly longer; think of the Scottish Blackface, Lonk, Derbyshire Gritstone, White-faced Woodland (the Woodie) and Swaledale. The oddity in this category is the Woodie, which has a very high-quality fleece, due to the fact that in the 16th century, the king of Spain overruled a Spanish law forbidding the export of the Merino (which originates in Spain, not Australia) and gifted some quality sheep to George III (nicknamed Farmer George). George III then gave some to the Duke of Devonshire in Derbyshire who crossed them successfully with the Woodland Vale sheep of South Yorkshire and North Derbyshire, which were typical small hill sheep, resulting in the Woodie, which is arguably the biggest of hill breeds and has a high-quality fleece. At the same time, according to diaries recorded by Daniel Defoe (author of *Robinson Crusoe*), the Merino was crossed with the wool-free Wiltshire Horn, but this crossing was not successful.

The Wiltshire Horn is part of the make-up of the modern breed, the Easy Care sheep. As it is free of wool, merely having a 'hair' coat, which it sheds, there is no need for shearing. There are many advantages to this: shearing contractors can be hard to find; the value of wool may not cover the cost of shearing; there is no 'sweaty' wool for flies to lay their eggs in that hatch into maggots and eat the sheep alive; and there is no long wool at the rear for the wet faeces of lush spring and autumn grass to become attached to, with the resultant task of dagging (see Glossary, page 92).

Hill sheep were bred pure to maintain the flocks 'hefted' to their particular mountain or moor. They were also bred to a bigger-framed sheep to produce larger, more prolific crossbred ewes for the lowlands, where the grass was more plentiful. The crossbred ewe, being bred to a meaty Down ram, could produce lambs to be either fattened on grass on their farm of origin or, more traditionally, sold to arable farmers to fatten on forage crops sown after the cereal harvest, or perhaps to the Shire farmers to fatten on autumn grass. These crossbreed ewes include the Greyface (Border Leicester ram on Scottish Blackface); English Mule (Blue-faced Leicester on Clun); Welsh Mule (Blue-faced Leicester on Welsh Mountain); Mule (Blue-faced Leicester on Swaledale); Masham (Teeswater or, traditionally, Wensleydale on Swaledale). The Down rams used to sire the lambs they bred would traditionally be South Down, Oxford Down, Hampshire, Shropshire and Suffolk, but in the 1960s and 1970s the French Texel and the smaller but very heavily muscled Belgian Beltex, Charollais, Ile de France, Berrichon du Cher and others were introduced with varying degrees of success.

DOWNLAND BREEDS

Downland breeds are traditionally bigger, heavier sheep, because they were developed on the rolling pastures of the Midlands and the Shire counties. Examples are South Down (small, solid and chunky); Oxford (slow to mature and very large); Suffolk (large, quite quick to mature and well-fleshed); and Shropshire (large, heavily fleshed and quick to mature). Border Leicesters are big, framy sheep, more suitable for crossbreeding to produce a prolific ewe than as a terminal sire (rams from large, meaty breeds of sheep, which will produce a meaty lamb no matter how slim the ewe it is crossed with). Wensleydale and the Cotswold 'Lion', both huge sheep with dreadlock wool, are slow to mature but with superb meat.

OTHER INTRODUCTIONS

As well as the imports of terminal sires in the 1960s and 1970s, other breeds were also introduced. The Bleu du Maine and Rouge de l'Ouest are big-framed, milky, prolific sheep, very good for producing crossbred ewes. The Bleu du Maine crossed with a Lleyn ewe (arguably one of the most prolific native breeds), for example, is a wonderful ewe.

The Finn was introduced in the late 1960s, and was popular for a while in producing the Finn cross Dorset, for very early lambing; however, too many of the crossbreed ewes had three lambs, which was found to be impractical. (Most sheep, like the Finn, only have two functional teats; by the time the two strongest lambs have had their fill there may not be any milk left for the third, who will require high levels of attention, through bottle feeding or fostering, in order to survive.)

The Friesland is a much bigger sheep than the Finn. Prolific and very milky, it was involved in the development of the British Milksheep. As the name suggests, the breed has been used on sheep dairy farms.

BREEDING TIME

The breed of sheep you select affects, to an extent, when the lambs will be born. Geographical effect on breeding season is quite marked in that one can relate the time of year when a sheep naturally breeds to its region of origin. For example, the Downland breeds will naturally breed early in the season. The South Down originates from a region where spring comes in late February and winters can be relatively mild. Consequently they will naturally accept the ram in late July and August, to lamb around Christmas. The hill ewes, such as Welsh Mountain, will only accept the ram in late autumn. They naturally produce the lambs in late April, when the warmer weather and spring grass appear, with sunshine for the lambs and plenty of fresh grass for the production of milk.

The onset of the oestrus cycle is brought about by the reaction of the sheep's pituitary gland to day length; only a slight lessening of daylight will induce Downland breeds to come in season, whereas it takes a much shorter day length to induce the hill breeds to become receptive to the ram. It is possible to control oestrus by the use of intra-vaginal hormone sponges if it is important to align lambing time with time off work. It is also possible to induce oestrus by only allowing restricted daylight hours in midsummer. I put 20 ewes on a system of nine hours of daylight in July and they all lambed within seven days, two weeks before Christmas.

THE BREEDS:
A SMALL SELECTION

*There are so many breeds of sheep that one can almost
certainly find one of a size, shape, colour and temperament
to suit one's needs.*

The following list cannot hope to be exhaustive, but I
hope it will give you an idea of what is available and what
might suit your needs. There is no suggestion that the breeds
listed are of superior merit to those not appearing in the list,
rather they are those with which I have worked during my
years handling sheep. Furthermore, wherever you live in
Britain it is also possible to keep a 'local' breed of sheep.

First are listed, in alphabetical order, the breeds usually
associated with the expression 'terminal sire'. In other words
the progeny of these rams are used in commercial flocks to
sire lambs destined for the table.

TERMINAL SIRE BREEDS

BELTEX

Derived from the Texel (see page 28), this breed was developed
in Belgium. The crossbreed lambs have a very high yield of
meat, with a high killing-out percentage (the proportion of
saleable meat on a lamb once it reaches the butcher's shop).

DESCRIPTION: Smaller than the Texel, on shorter legs, the
huge double muscles on the hind quarters give the Beltex
a wedge-shaped appearance.

CHAROLLAIS

Originating in France. This breed is excellent as a terminal sire in that they combine very good muscle structure with a fine bone, have a good meat to bone ratio, and a fine shoulder, which results in easy lambing. Prolific for a pure breed, a 200 lambing percentage being quite common in flocks of mature ewes.

DESCRIPTION: A large sheep with a pink head and legs that are free from wool (see colour section for illustration).

DORSET DOWN

Can be used as a sire on any breed of sheep, even ewe lambs, and be fairly sure of an easy lambing. It will survive in a range of climate conditions and will breed at any time of year.

DESCRIPTION: A medium-sized sheep, with excellent muscling of the hind quarters and loin, with a fine head and shoulders (see below and colour section for illustration).

Dorset Horn and Polled Dorset

Dorset Horn and Polled Dorset sheep can breed at any time of year and therefore the rams are useful as terminal sires to produce lambs in early lambing commercial flocks. The lambing percentage is about 160 per cent and they are often used to produce a crossbreed ewe that will lamb early in the season.

DESCRIPTION: A medium-sized sheep, white-faced with a remarkably white fleece and quite well-fleshed on a solid 'square' frame.

Hampshire

The Hampshire will mature quickly and early – a good characteristic for those who wish to sell the lambs early. Lambing percentage is around 160 per cent. They may look slightly smaller than the Shropshire (see page 27), but due to the very heavy muscling will weigh about the same.

DESCRIPTION: A solid, heavily muscled sheep with a dark brown, nearly black face and wool coming down from the poll (top of the head) onto the face (see colour section for illustration).

OXFORD DOWN

Our biggest native Downland terminal sire breed. The
Oxford Down has been developed in recent years to the
point that that, if fed correctly, its fast growth rate can
result in crossbreed lambs being fit for slaughter at a very
early age. If allowed to mature more naturally, it will result
in a much heavier carcase of 20–25kg (44–55lb), which has
not gone too fatty.

DESCRIPTION: Traditionally a big, raw-boned sheep. It
is not as heavily muscled as the Hampshire (see page 26),
but in recent years the degree of muscle has improved
dramatically. A dark chocolate-brown face, free of wool,
but with wool on the poll.

SHROPSHIRE

The Shropshire lays claim to the world's oldest sheep
society with the first flock book being published in 1882.
Long sheep, which are very long-lived, finish well on grass,
thrive in a large range of climates. They will graze orchards
and plantations without damaging the trees. Lambing
percentage about 160 per cent.

DESCRIPTION: Pink skin, off-brown head and a relatively clean
face with wool over the poll and some wool on the legs (see
colour section for illustration).

South Down

Because of its relatively small size, the ram can be used to sire crossbreed lambs, which are ready for slaughter at a very early age. The breed can achieve 200 per cent lambing with mature ewes.

DESCRIPTION: Small and chunky, like a concrete block on legs, South Downs have a 'mousy brown' face and legs with short ears covered with wool.

Suffolk

No list of British breeds of sheep is complete without a reference to the Suffolk, the most popular and commonly used terminal sire in the UK.

DESCRIPTION: Naturally polled, black head and legs with no wool, short fleece and large black floppy ears (see colour section for illustration).

Texel

A medium-sized sheep, justifiably famous for its carcase quality. It is used very extensively as a terminal sire because it imparts its phenomenal carcase qualities to its progeny. A butcher's dream, which is why Texel cross lambs usually win the lamb carcase competition at shows.

DESCRIPTION: A chunky solid head covered in short white hair, with a black nose, white legs and black feet (see colour section for illustration).

NON-TERMINAL SIRE BREEDS

BLEU DU MAINE

Introduced in the 1980s from France, the Bleu du Maine has become successful as the sire of a prolific and milky crossbred ewe with a superb conformation (body shape).

DESCRIPTION: A large sheep with clean face and legs, and a blue-grey colour. No horns, tight fleece, white wool, with a good muscle shape (see colour section for illustration).

BLUE-FACED LEICESTER

Sire of Mules (see page 37). Descended from the Dishley Leicester and developed by Robert Bakewell in the 18th century. Due to the preferences of Northumberland sheep owners Blue-faced Leicesters have characteristics that differ from the Border Leicester (see page 30). In recent decades the Blue-faced Leicester has become broader and this has resulted in the Mule becoming a larger sheep, which has made a very strong contribution to the continued success of the Mule. They have very little wool, so the Blue-faced Leicester is a shearer's dream. Prolific, with plenty of milk.

DESCRIPTION: The face and skin have a bluish tinge. The nose is not so aquiline as the Border Leicester and the ears are not so hugely prominent and erect, but nevertheless a very aristocratic sheep (see colour section for illustration).

BORDER LEICESTER

Very distinguished sheep, a flock of which is a sight not to be missed. This breed is directly descended from the Dishley Leicester, developed by Robert Bakewell in the 18th century.

DESCRIPTION: A large raw-boned sheep, with a very aristocratic nose and large erect ears, on a head held very erect (see colour section for illustration).

CAMBRIDGE

A synthetic breed, so-named because it was developed by a number of breeders and researchers at Cambridge University, primarily John Owen and Alun Davies. Many mature breeds were used in the development of the Cambridge, particularly the Clun Forest (see page 30), and the females used had to have bred triplet lambs for a number of successive years. They produce prolific crossbreed ewes, with the milk to feed triplets.

DESCRIPTION: The sheep is medium-sized, with a relatively fine-boned, square frame, not heavily muscled. The face and legs are pale brown to chocolate brown and are free of wool.

CLUN FOREST

A mature Welsh breed from the Border Counties, based
upon the area it is named after, the Clun Forest. It developed
from a number of mature breeds, including the Welsh and the
Shropshire (see page 27), and others now extinct.

DESCRIPTION: Pale brown, medium-sized sheep with a very
tight fleece. There is wool on the poll, but a clean face. The
Clun Forest has sharp and erect ears, giving the impression
of a very alert sheep.

COTSWOLD

A large sheep. The wool is long and of high quality and
both male and female are polled (free of horns). These
sheep, affectionately known as the Cotswold Lion, are
descended from sheep that were found in the Cotswolds
during Roman times. The sale of their wool during the
Middle Ages resulted in many a wool merchant making
his fortune, the legacy of which are the large churches to be
found in Cotswold villages and towns. They were involved in
the development of the Oxford Down breed in England (see
page 27), and the Oldenburg in Germany. It was very popular
as a producer of meat but, as demand for large joints of meat
fell due to economic decline of the early 20th century, so
did the popularity of the Cotswold. They will produce a
heavy fleece and average about 175 per cent lambing as
mature ewes.

DESCRIPTION: A large and wonderful character sheep, with
white face, black nose and 'forelock' of wool (see colour
section for illustration).

DERBYSHIRE GRITSTONE

Originating in the hills on the edge of the Derbyshire Peak District, this breed is capable of producing 150 per cent lambing on the hill. It is a good cross with the Down rams and has recently become quite popular for crossing with other hill ewes to produce a hornless crossbreed. The advantage of this is to eradicate the problem of headfly – similar to fly strike (see page 57) – that is associated with horned sheep.

DESCRIPTION: A hornless, strong, very attractive sheep with a very light fleece.

FINNISH LANDRACE

The Finnish Landrace was particularly popular in the late 1960s and early 1970s when it was used to produce the Finn Dorset, which had the size and early lambing facility of the Dorset with the (diluted) prolificacy of the Finn. It is a very prolific breed indeed.

DESCRIPTION: Medium-sized, fine-boned sheep with a clean face and white legs (see colour section for illustration).

FRIESLAND

Used as dairy sheep due to their high milk yields and, being very prolific, they produce easy-lambing, milky, prolific crossbred ewes, which will lamb early in the season. They are not very common as a crossbreed, but if a terminal sire that is extremely heavily muscled – such as a Beltex – is used, they can be very successful indeed.

DESCRIPTION: A large open-framed sheep with a relatively loose fleece, white in colour with pink nose and a large, roomy pelvis, which results in easy lambing (see colour section for illustration).

GREYFACE

At one time this was possibly the most numerous sheep in Britain, the result of crossing the Border Leicester ram (see page 30) on a Scottish Blackface ewe (see page 39). A very successful cross, the ewe has mothering ability and hardiness from the Scottish Blackface and milk, frame and prolificacy from the Border Leicester.

DESCRIPTION: A medium-to-large ewe with a slightly aquiline nose from its sire. It has a mottled greyish face and legs.

HEBRIDEAN

Supposedly introduced to the British Isles by the Vikings, these are very easy sheep to keep; the wool is easy to spin and a Hebridean wool hat is very warm at lambing time. Although small, the lamb produced is absolutely superb with a unique texture and flavour. These sheep thrive in an outdoor environment, needing minimum shelter and will lamb very easily whether bred pure or crossed to a terminal sire breed. Thriving in large or small flocks, 175 per cent lambing is no problem on the lowlands and foot problems are rare.

DESCRIPTION: Small, horned, black sheep with clean face and legs (see colour section for illustration).

HERDWICK

Arguably Britain's hardiest of breeds; not very prolific in its natural environment, but more so in a lowland environment. They have proved very successful as a 'hobby' sheep and cross well with terminal sires.

DESCRIPTION: White face and legs with worn grey wool. The ewes are hornless and the rams usually horned. The lambs have wool that is almost black when born. They have large feet, which suits their natural habitat, the Lakeland Fells (see colour section for illustration).

KERRY HILL

The Kerry Hill originates from the Welsh Borders and gets its name from the village of Kerry. Originating on the hill, it is equally at home in large or small flocks in the lowlands and it will cross well with a Down sire to produce a quality lamb. This is a classic example of a breed's changing fortunes – in the mid- to late 1960s when 19,000 sheep would be sold in a day at the Craven Arms sheep sales, 7,000 of them would be Kerry ewes. Their popularity fell to such an extent that by the mid-1990s they qualified for the rare breed sale. Thankfully they are staging a revival.

DESCRIPTION: Attractive, medium-sized sheep with a dense fleece and very eye-catching black markings on the face and legs. Black, erect ears (see colour section for illustration).

LEICESTER LONGWOOL

The wool of the Leicester Longwool is very long and the fleece yield can be up to twice that of other breeds. The wool is extremely popular with home spinners. Ewes will produce a wonderful lamb for the table, either as a pure-bred wether or as a crossbreed from a terminal sire.

DESCRIPTION: Huge 'dreadlocked' sheep, the face covered with white hair, but a bluish tinge to the ears and nose (see colour section for illustration).

LINCOLN LONGWOOL

The Lincoln is the largest of the Longwool breeds, with a large heavily boned frame, carrying probably the heaviest fleece of any breed. Although not the most prolific of breeds it will produce large carcase lambs at a relatively early age. A very easily managed sheep with a good temperament, but remember that a large ewe, even though she may be good-natured, can still knock you over if she decides to wander from point A to point B and you are in the way.

DESCRIPTION: Hornless and white-headed with dark ears, the face is covered with long strands of wool growing down from the forehead (see colour section for illustration).

Lleyn

Originating on the Lleyn Peninsula in North Wales, this sheep is classed as a Lowland breed. They are very prolific and when well fed and on lowland pasture, a lambing percentage well in excess of 200 per cent is possible with mature ewes. A personal favourite in the long list of sheep breeds.

DESCRIPTION: Free of horns, with a white face; more solidly built than the apparently similar Welsh Mountain ewe.

Lonk

One of the largest British hill breeds, the Lonk has a good tight fleece. Relatively prolific, they are traditionally based upon the Lancashire and Yorkshire Pennines. Crossed with Down rams they can produce quality lamb straight from the hill.

DESCRIPTION: Both sexes are horned; face and legs are marked with black and white.

Manx Loghtan (or Loaghtan)

Derived from Viking introductions and isolated on the Isle of Man, hence the name. Very similar to the Hebridean (see page 33) in character and performance.

DESCRIPTION: Mousy brown (*loghtan* in the Manx language) colour. Both sexes have horns. Like the Hebridean, this breed can have two, four or six horns, occasionally polled (see colour section for illustration).

Charollais

Dorset Down

Hampshire

Shropshire

Suffolk

Texel

Bleu du Maine

Blue-faced Leicester

Border Leicester

Cotswold

Friesland

Hebridean

Illustrations are not to scale.

Herdwick

Kerry Hill

Leicester Longwool

Lincoln Longwool

Manx Loghtan
(or Loaghtan)

Masham

Illustrations are not to scale.

Scottish Blackface

Soay

Swaledale

Wensleydale

White-faced Woodland

Wiltshire Horn

MASHAM

The traditional Masham is a medium-sized ewe, crossbred, being a Wensleydale on a Dalesbred or Swaledale, but is now sired by Teeswater rams. They are long-lived, hardy, milky and prolific, 200 per cent lambing or more being no problem in mature ewes.

DESCRIPTION: A hornless breed, the Masham has a black nose on a black head with a white stripe. The wool is rather like dreadlocks (see colour section for illustration).

MULE

Any crossbreed sired by a Blue-faced Leicester (see page 29) is known as a Mule. They are a commercial sheep farmer's dream; the sheep shearers love their clean head, neck, belly and legs, which is soon sheared. The fact that the Blue-faced Leicester imparts a degree of uniformity, regardless of what ewe it is used on, means that the lambs produced by a Mule, usually sired by a Suffolk, Texel or Charollais, will be fairly uniform in size, which is ideal from the supermarkets' point of view; standard-sized, easily packaged, consistent supply of material. The most common strain of Mule is a Blue-faced Leicester on a Swaledale. Other popular combinations are Blue-faced Leicester on a Clun (English Mule); North Country Cheviot (North Country Mule); or the Welsh Hill ewes (Welsh Mule). The Blue-faced Leicester provides frame, milk and prolificacy.

DESCRIPTION: The traditional mule is pale brown to grey. It is speckled on its muzzle and has erect, speckled ears. Due to the variety of 'mules' produced, there are many different forms.

Rouge de L'Ouest

Imported in the 1980s from France. Medium in size with a fine bone and very good muscle development, they are ideal for use on ewe hoggs due to the fine bone. Because they have been developed for their lean meat, the breeder has a larger window of time during which the lamb can be sold without becoming too fat. Very milky and prolific.

DESCRIPTION: Short, dense fleece. Hornless sheep with a clean head and legs, which are pink to a deep pastel brick red in colour.

Scotch Half Breed

The result of using a Border Leicester ram on a North Country Cheviot ewe. Milky, prolific and solid. The use of a Texel ram on Scotch Half Breed ewes produces an excellent crossbreed lamb for the supermarket.

DESCRIPTION: Lovely white sheep with a dense fleece. The Scotch Half Breed is fine-boned and very square in shape.

SCOTTISH BLACKFACE

A very hardy ewe developed to thrive on the harsh moorlands of Scotland. Crossed with the Blue-faced Leicester (see page 29), the result is the hardy, milky and prolific Scottish Mule; crossed with the Border Leicester (see page 30), the result is the Scottish Greyface, again hardy, milky and prolific. Both these crosses are successful on lowland farms in the production of prime lamb when crossed with terminal sires.

DESCRIPTION: Both male and female are hoofed, usually with a white poll and grey-haired nose, black feet and bracken-coloured white legs. A long coarse fleece gives protection from the heavy rain and cold in its native environment, the Scottish hills and mountains (see colour section for illustration).

SOAY

The original 'primitive' sheep of the British Isles. They will settle well in small flocks, but fencing must be of the highest quality as they can, if put under stress, clear a fence well over 1m (3ft) high. The small joints of lamb or mutton have a gamey flavour and are wonderful when accompanied with a glass of red wine.

DESCRIPTION: Small and horned, with short self-shedding chocolate-coloured or fawn wool. The mature males can develop a distinctive mane (see colour section for illustration).

SWALEDALE

A very hardy breed. The Swaledale may be difficult to keep at home at first as they have for generations roamed over vast areas of moorland. However, once they have bred a crop of lambs, they will probably settle down and, after a couple of years, become 'hefted' to your property. They have been developed by breeders since the 19th century, and probably earlier, to be very hardy and able to survive where few other breeds of sheep can. This hardiness results in our exposed moorlands being grazed and maintaining the wonderful scenery we expect in Britain. It is possibly the most influential female in the British sheep industry in that, crossed with the Teeswater and to a lesser extent the Wensleydale, they produce the Masham (see page 37); crossed with the Blue-faced Leicester they produce the Mule (see page 37).

DESCRIPTION: Horned, with a thick, dense fleece. Swaledales have a dark head with a nearly grey muzzle, the dark head becoming grey with age (see below and colour section for illustration).

TEESWATER

Originating in Teesdale, they were not very numerous until
their crossing ability was discovered in the early 20th century,
producing the Masham when crossed with the Swaledale.
They are long-lived and very prolific and milky; triplets and
quads being common in mature ewes when well managed.

DESCRIPTION: Large sheep with a light, curly fleece and
no horns. The face and head can be off-white to blue/grey,
with a dark brown nose, eye surround and ears.

WENSLEYDALE

A beautiful breed of sheep, developed in the Wensleydale
region, mainly for crossing on hill ewes to increase their size
and milking ability while retaining the hardiness of the female.
The breed is unique in that it can trace its origins to one ram,
Bluecap, in the mid-19th century. He was famous for his blue
colour, long wool and enormous size, which he passed on
to his progeny, such was the predominance of his genetic
characteristics. The Wensleydale ram on a Swaledale ewe
was the original Masham.

DESCRIPTION: A very large 'dreadlocked' Longwool, polled in
both sexes, with a distinctive velvety blue face and ears (see
colour section for illustration).

WHITE-FACED WOODLAND

A mature Derbyshire breed developed from the Penistone hill sheep of South Yorkshire. The high-quality wool, which is characteristic of the Woodie, is the result of a gift by King Charles of Spain to King George III of England of some Merino rams, which ended up in part in the care of the Duke of Devonshire at Chatsworth. These were crossed very successfully with the Penistone ewes to develop the White-faced Woodland – the 'woodland' part of the name coming from the 'woodland' vale of North Derbyshire where the original Penistone was also to be found.

DESCRIPTION: Horned in both sexes, heavily curled in the male with a muscled tail that is left in the male but docked relatively long in the female. Woodies are white, with a pink nose and clean face and legs (see colour section for illustration).

WILTSHIRE HORN

The only British wool-free sheep. Very easy to care for, in that fly strike is no problem. It crosses very well with the Down breeds (such as Dorset, South and Oxford).

DESCRIPTION: Both males and females have horns. The sheep are white, fine-boned and medium in size (see colour section for illustration).

BUYING YOUR SHEEP

*Having worked so hard to prepare for the
arrival of your chosen breed, buying your
sheep is second only to lambing in the
excitement ratings.*

Once you have decided to go ahead and buy some sheep
it is most important to take your time and do plenty
of research. Check the internet for lists of breed sales, and
auctioneers will also list their sales. Farming journals and
magazines will also list forthcoming sales.

CATEGORIES OF SALES

There are basically four categories of sales: commercial sales,
held in the autumn, when farmers and breeders sell their
surplus breeding stock; breed society sales, when pedigree
breeders sell breeding males and females and buy in new
blood lines; 'rare breed' (and I use this term in its broadest
possible sense) sales, where the less popular breeds are sold;
and, finally, private sales.

All types of public sale are worth a visit, just to see what
is available. They begin in late July or early August often
on Saturdays and, if you plan to visit a different type of sale
each weekend through to early or mid-September, you will
still have time to visit the late September and October sales
to buy your sheep and put them to the ram in late autumn
to lamb in spring. Hopefully the weather will be on your
side, with the sun shining on newly born lambs.

All the sheep at commercial sales will be hardy, in that they all come from commercial working farms. Sheep at commercial sales will usually be sold through an auction ring, the auction staff bringing the sheep from the pens to the ring, with the bidders gathered around the ring. Sheep are sold in groups; sometimes as few as five, but more usually ten, 20, 30 or even 50 in one group! Most auctions have catalogues available, which may be a rudimentary typed list. This catalogue will also give the health status – MV-accredited or not (for more information about Maedi Visna, see page 60) – and the level of vaccination, worming and dipping of all sheep sold.

Stand and watch at a few auctions to get the hang of what is going on. Each auctioneer has his own technique and some even seem to have their own language. Watch the people bidding. Some wave a catalogue in the air, some shout a bid, some nod, others raise a hand, some a finger, some wink, some seem to do nothing, but still buy the sheep; and by watching you will learn how much you can expect to pay. When you are looking for your sheep, do not be afraid to ask for advice from the older farmers; they'll have probably forgotten more about sheep than you will ever know.

When choosing sheep, look at the pen of sheep as a whole – do the animals look evenly matched? Look at their teeth – they should rest neatly against the dental pad of the upper jaw (see diagram). Remember, sheep have teeth at the top and bottom towards the rear of the jaw, but only on the bottom jaw at the front. The best sheep to buy in some respects are shearlings – namely those that are 15–18 months old at the

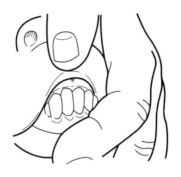

time of the autumn sales, having been sheared once. However, they will probably not have lambed before, which could result in both you and your ewe having big 'L' plates on at lambing time. A shearling ewe will have two, possibly four, broad teeth and her milk teeth will be fading away.

A 'two shear' ewe will be catalogued as such, but always feel the udder of a sheep that has had lambs; it should be soft, with no lumps. If the udder is lumpy in any way, shape or form, no matter what the vendor says, do not consider buying the sheep. Equally, if a two shear or older ewe looks very fit, fat and healthy compared with the others in the group, feel the udder. If there is no udder (and remember it will be small because she is not lactating – compare it with that of the others in the group), the ewe probably has never bred at all and never will.

When the sheep run from the pens to the ring, try to gain a vantage point at the ring where you can see the entry alleyway and the ring – watch how they walk. Any sheep with a lameness problem will be easily visible.

Finally, always be sure of what you are buying. A non-farming neighbour of mine acquired a small farm. He couldn't understand why the 20 beautiful young sheep he had bought, after spending the winter with his newly purchased prize ram, had not produced any lambs. He was embarrassed to be told that he, a doctor, had purchased 20 wethers – castrated males.

BREED SALES

Sheep at a breed society sale will be 'hardy', as at a commercial auction. At a breed sale all sheep will be of the same breed or an amalgamation of breeds if a number of societies are holding a joint sale.

The sheep are usually sold in guineas (£1.05) – an old tradition (and long may these British peculiarities continue). The price the auctioneer sells at is the price per animal in the ring, at all sales, unless otherwise stated, and will probably be much higher for pedigree registered stock at a breed sale.

The most important factor to beware of at breed sales – in addition to those mentioned on pages 44–45 – is that there will be people at breed sales who are selling sheep from a very small flock, where the sheep lead a very sheltered life. It may be prudent to ascertain in the course of conversation how many ewes a breeder has and look to see if the sheep have buckets full of concentrate feed in their pen. If they are fit, well fleshed and healthy with a rack full of hay/haylage and look happy and content, with no sign of tubs of concentrate, the sheep will probably be well able to fend for themselves, living primarily off grass as sheep should.

RARE BREED SALES

Having bred a large number of pure breeds, my favourite
sales are the rare breed sales. There is usually a wonderful
atmosphere. Friday is usually show day when, after vet and
breed inspections, the animals there for show as well as sale
enter their respective classes. Some of the breeds are card
graded by breed society inspectors – red for exhibition quality,
blue for a good breeding sheep and yellow for average. The
card grading system is a useful aid for the novice buyer. Show
day offers a good opportunity to look through the multitude
of breeds and confirm your decision.

Saturday is usually sale day at rare breed sales, no matter
how big or small. Sometimes the sheep are sold in the pens
with auctioneer with buyers moving from pen to pen. As
mentioned earlier, while the price is per animal (in guineas),
beware, because the entire contents of the pen may be sold –
rarely more than three at a rare breed sale – or they may be
selling one animal.

PRIVATE SALE

The fourth method of buying stock is to buy privately. Most
breed societies have helpful sales officers who will put you
in touch with people who have stock for sale. Before buying
stock privately it is a very good idea to attend a few public
sales to help you form an opinion as to what prices you
might be expected to pay.

However, the most important thing to be sure of is that your
sheep look sharp, healthy and alert, with a good fleece, clear

eye and sound feet. Always check that a sheep is standing on its toes, not down at the pasterns (the heel part of the sheep's foot). A sheep's feet are akin to our fingers; when walking on their toes they are walking on hard, keratinous, horny tissue, protecting the softer flesh further up the foot. Dropped pasterns will expose the softer tissue to the mud and wet of the land in a typical winter and increase the likelihood of foot infections.

Moving up to the teeth: sheep eat in a totally different manner to cows. Cows, although they have the same dentition as sheep, wrap their tongues around the grass and tear it off against their bottom teeth, eating virtually anything they encounter. Sheep are far more delicate, 'nibbling' with their lips, 'feeling' the grass before they grip it between their bottom teeth and dental pad of the top jaw and then tear it off. Hence, uniform teeth, correctly matching up to the dental pad of the top jaw are very important (see diagram, page 45).

Finally, if the sheep of your choice are pedigree, ensure they conform to the breed standard. A description will be available from the sales officer of the society.

GENERAL HEALTHCARE

Although there is a legal requirement to alert Defra to notifiable diseases if they occur, there is no legal requirement for any vaccination. If at any time you are unsure of the health of your sheep, seek veterinary advice.

Maedi visna, scrapie, blue tongue, fly strike, liver fluke, worm infestation (nematodirus), listeria, hypocalcaemia, toxoplasmosis, enzootic abortion, septicaemia, foot rot, sheep scab, pasteurellosis and coccidiosis: these ailments are listed in no particular order but you can quickly see why shepherds say that the sheep is the only animal on earth that is looking for the quickest way to die.

Whatever sheep you purchase, always ascertain its vaccination status, which should be stated in the sale catalogue. Most sheep are vaccinated with a 'pasteurella' vaccine, of which there are two major brands available. Your vet or local farm supplier will be able to advise you of the best one to use. These vaccines protect against seven or eight of the most frequent forms of clostridial infection, pasteurellosis being the most common. Initially, two vaccinations are required, three to four weeks apart and thereafter an annual booster shot. With sheep you have purchased, it is quite acceptable to administer the booster shot early to facilitate fitting in with your system; do not administer late as protection will fail.

Your home-bred stock, if they are to be sold for slaughter, may not need to be vaccinated. However, animals that you wish to keep for breeding certainly do. The smallest container of these vaccines you can purchase is 100ml, which is enough for ten sheep. It has a short shelf life once opened, so you will not be able to store it for later use.

The following list includes diseases and conditions you may come across as a keeper of sheep, but please bear in mind that this book is not a substitute for advice from a professional. If you are in any doubt about the health of your sheep you should seek veterinary advice.

COCCIDIOSIS

This is a parasitic condition that usually affects lambs and is recognisable as a grey scour (diarrhoea) with the lambs rapidly losing condition. It is not usually a problem in mature ewes that have met the problem and come to terms with it. It usually becomes apparent at the five- to six-week stage, when lambs begin to obtain a greater proportion of their feed from pasture as opposed to their mother's milk. Treatment is by 'drenching' the lambs once with an appropriate product, or daily sulphonamide injections for a few days. If in doubt, your farm supplier will show you how to assemble the drenching gun.

SHEEP SCAB

This is a notifiable disease and, if you spot the symptoms of this condition, it is a legal requirement that you notify Defra immediately. Sheep scab is caused by a mite settling on the skin of the animal, where it causes irritation. This causes the sheep to rub against anything to hand and the wool will soon rub off the sheep and on the floor. The affected sheep will develop raw sores and non-infected sheep can become infected by contact with contaminated sheep wool or by sharing the same rubbing post. Contact your vet immediately. He will probably take a skin sample to confirm, under the microscope, your worst fears. An injection will usually cure the problem, but severe cases may need to be culled on welfare grounds.

FOOT ROT

There are two causes of foot rot. The first is infection, due to bacteria causing the tissue to rot and the other is by mud trapped in overgrown or undergrown toenails becoming warm and softening the hoof, eventually causing it to rot. In both cases, trim back all the necrotic tissue with foot trimming pliers, apply antiseptic aerosol spray and turn the sheep onto (preferably) clean pasture. As a vet told me 40 years ago, when trimming feet you need a sharp knife and a stout heart.

A formalin or perhaps an alternative, less aggressive footbath may be an option if your flock is prone to foot problems. A footbath is a shallow trough, specifically designed for the purpose, which is placed in the race (the gated one-sheep passageway in which to restrain sheep). Your sheep, ideally with clean feet, stand in the footbath for a period of time (dependent on the product you are using), then move onto clean, dry hardcore or concrete, before moving to clean pasture.

SEPTICAEMIA

You do not want to see septicaemia in a sheep, because it is a very serious condition usually resulting in death. There are three main causes. If a sheep has had a difficult lambing, she may have suffered internal bruising. If in any doubt, treat the ewe with an appropriate antibiotic, which you obtain from your vet. The vet will inject your sheep for you, but the sooner you learn to do it yourself the better. If you do not administer antibiotics and there is a problem, about 24 hours after lambing the ewe will become dull and listless and her body temperature will rise as the infection takes over. The second cause can also occur during lambing – although it is rare, a ewe may retain her placenta. If this does happen, again treat with an antibiotic. The third, most common cause of septicaemia is due to udder infection. If on your daily inspection you come across a sheep, perhaps a bit dull, standing on her own and not grazing or maybe walking a little stiffly, catch your sheep (not always easy), and examine the udder. The udder should be soft. If it is solid, hard or lumpy, there is an infection. If infected, treat the ewe with an antibiotic injection and empty the affected part of the udder by 'hand milking' the ewe. It will certainly smell, the ewe will object and it probably will not milk again on that teat but at least you still have the ewe – plus a few bruises. It is certainly a good idea to sell such sheep before next year as a cull ewe; she will be sold for mutton and the look of the teat will not be a problem. If they only have one functional teat they will almost certainly have two or three lambs at next lambing – that's Murphy's Law – and you will have to foster the lambs or bottle rear them, which is not to be recommended.

ENZOOTIC ABORTION

This is an infection that is usually 'bought in' so always check the health status of your sheep before you buy. If lots of ewes lamb prematurely, consult your vet. A vaccine is available.

TOXOPLASMOSIS

NEVER allow a girl or woman who may at some stage in her life have children to handle a wet, newly born lamb without plastic gloves. Females must at all times pay very careful attention to washing their hands. Why? Women of child-bearing age, if handling an apparently normal lamb that may have come from an affected sheep, can become infected with toxoplasmosis. Toxoplasmosis is caused by an infection carried by rats. The rats defecate on sheep feed, or are eaten by cats, which can then equally infect sheep feed, and the sheep eat the contaminated feed. Toxoplasmosis eggs develop in the ewe's gut, migrate to the placenta and cause the ewe to abort. Once the ewe has been infected and aborted, she will usually breed normally in subsequent years.

A fairly effective control is to include a coccidiostat in the ewe's feed, which kills the parasites in the gut before they become a problem. Although we do not have this problem on our farm as far as we are aware, we include a coccidiostat in all pure-based concentrate for our in-lamb ewes.

HYPOCALCAEMIA, HYPOGLYCAEMIA AND PREGNANCY TOXAEMIA

These problems are caused by a shortfall in the nutrient status of a pregnant ewe. Hypocalcaemia is a calcium shortage in the blood supply, caused by the nutritional demands of the foetus (or more probably two, three or four of them) being far in excess of what the sheep's diet is supplying. Hypoglycaemia is an energy deficiency caused by an imbalance between foetal demands and dietary intake. Pregnancy toxaemia is a term used to describe the dull and listless appearance of a ewe that is hypocalcaemic or hypoglycaemic. She will appear weak all over (as opposed to listeria, which only affects one side), perhaps vaguely unaware of her surroundings and disinterested in feeding.

To treat such a sheep, first of all place her in isolation. If you have recognised the symptoms early enough, because she is on her own she will feed herself, whereas left in the group she probably would not, thus exacerbating the problem. Rather than plain water in the bucket, place electrolyte solution in her drinking water, rich in energy and essential salts. These are available from any farm supplier.

If you fear the case is more advanced, drench the sheep with electrolyte, but be very careful. It is very easy to drown a sheep when you drench it. Injections of calcium borogluconate may also help; it is rich in CA^{2+} ions, which aid nervous/muscular coordination, and glucose for energy. This can be administered either subcutaneous (under the skin – S/C) or intravenous (I/V), but it would be better to get the vet in to perform the latter operation.

The best way to prevent hypocalcaemia and pregnancy toxaemia is to feed the ewes appropriately. If you can have your ewes scanned for multiple births, you will be able to feed the sheep accordingly. Some suggestions follow:

- For four or more lambs get the concentrate ration up to 1kg as quickly as possible, eight weeks before lambing
- If you have triplets, start at 100g six weeks before lambing and up to 1kg four weeks before lambing
- For twins, start at six weeks before lambing and build up to 750g one week before lambing
- In the case of just one lamb, if the hay is very leafy, a little feed (100g) about two weeks before lambing.

By adopting this feeding regime sheep carrying multiple lambs can build up sufficient reserves, in most cases, to carry them through the last couple of weeks when they may not have sufficient stomach space to eat enough to stay healthy.

LISTERIA

This is a condition brought about by contaminated feed. Hay or, perhaps more commonly, haylage that has been contaminated with soil from molehills is the usual scenario. It can also be caused by mouldy feed. The symptoms are primarily a degree of paralysis on one side of the body, affecting head, ears, eyes and legs. This results in the affected sheep walking round in circles, as opposed to hypocalcaemia (see above) where all the legs are affected.

Treatment can be successful, but will involve a long course of antibiotics plus a lot of hands-on care – primarily trying to assist the animal to drink and eat enough – until the drugs ease the symptoms and it is capable of drinking and feeding on its own.

WORM INFESTATION (NEMATODIRUS)

Adult sheep usually tolerate a low level of worm infection and do not need the same degree of treatment as lambs. We treat our adult sheep twice a year with a worm drench (anthelmintic). The first time should be around two to three weeks before they go to the ram, so that all their ingested feed is converted to benefit the sheep, not the parasite population. This will hopefully result in a ewe in prime condition with a good ovulation rate, ready for breeding. They are sent again when they ewe is penned with her lambs immediately after giving birth. For some reason a ewe will, immediately after lambing, show a large rise in the number of worm eggs she passes in her faeces (the egg count).

If the ewe is drenched with an anthelmintic (a drug that expels parasites), and is left penned up for 24 hours, she is only eating 'clean' hay and concentrate. She is not ingesting grass from a pasture contaminated with worm eggs so, due to treatment, she is clean. If she is then turned out onto a 'clean' pasture – with no other sheep having grazed it for at least three weeks – she will not be depositing vast quantities of worm eggs onto the pasture and the lambs will not be exposed to high levels of infection. However, unless such a three-week rotation can be practised, or your stocking levels are very low, the lambs will eventually need to be treated.

The signs of worm infestation are dirty rear ends, with accumulated wet faeces or scour on the fleece. You will notice that some of the lambs are beginning to look less robust. Warm, wet weather can increase the chances of worm infestation.

Although some schools of thought suggest routine dosing every three weeks, it is perhaps best to observe your flock and act when necessary, and usually only treat the lambs.

LIVER FLUKE

Liver fluke is carried by snails on grass in wet and marshy areas, and can be more of a problem in a wet summer. Once the sheep has eaten the snail carrying the fluke, the parasite eventually makes its way to the sheep's liver where eventually it will destroy the liver. Most worm drenches are also effective against liver fluke. If you do have a problem and have not noticed it by the time your flock go to the abattoir, the meat hygiene inspectors will inform you if there is fluke in the liver of your sheep and affected animals will be condemned.

FLY STRIKE

This problem is exactly as it is described – the sheep is attacked by flies. This occurs when the weather is warm and humid, and usually is found in the thick wool of sheep with a dense fleece, particularly the Down breeds. In most cases fly strike occurs on the back and upper shoulders, where under the fleece it can become very sweaty. The other favourite site for fly strike is the flanks, where faeces have accumulated either side of the tail in long wool. The flies lay eggs and very quickly these hatch into maggots, which begin to eat the sheep while it is still alive. Early signs are a slight darkening of the fleece due to the moisture produced by the maggots beneath. As the condition advances the fleece will fall off quite easily, exposing raw flesh where the sheep has rubbed. Vigilance is very important if you are to avoid infestation.

The treatment is usually effective, but not pleasant. You are advised to wear plastic or rubber gloves. With your hand clippers, remove the wool from around the affected area, leaving quite a good margin, about 10cm (2–3in), and then wash thoroughly with saline solution, rubbing the saline gently over the wound and more firmly into the surrounding wool. Spray the affected area with an antiseptic spray or perhaps smother it with a naturopathic ointment – a 'natural' antibiotic. For further advice, check with your vet.

You can obtain pour-on insecticides, which can be applied to the sheep's fleece to protect against fly strike, ideally around two weeks after shearing in mid- to late May. Contractors are also available to dip your sheep, totally immersing them in a mobile dip.

A further point to be aware of is the tendency for horned rams to suffer fly strike at the base of the horns. This is referred to as head fly. Although it is actually caused by a different fly, the result is the same. Infestation can be quite difficult to see, particularly in the darker-woolled breeds.

BLUE TONGUE

This is a notifiable disease. If you spot the symptoms of this condition, it is a legal requirement that you notify Defra immediately.

This is a midge-borne virus, of which there are many strains, and causes the face, nose and tongue of the sheep (and cattle) to become inflamed. This results in restriction of the airways, in effect semi-strangling the sheep. Oxygen deprivation causes the sheep to appear blue-tongued, hence the name of the disease. As she stands there, panting and salivating copiously, she is under a great deal of stress. Vet treatment may save the animal, but any stress factor, even pregnancy, may cause it to flare up again.

In the UK, a great team effort in 2008 by the NFU, Defra and the drug companies resulted in a vaccine being produced for the one strain that had arrived in the UK, carried by wind-borne midges from Europe. Sheep need an annual shot of 1ml, administered subcutaneously; it comes in 100ml vials and a special application device is also needed. Work is being undertaken continually to develop multi-vaccines to protect against all strains that may occur. Even if you have to buy 100ml for a small number of sheep, if it saves one pedigree ewe, it is worth it. A further tool in our attempts to keep this problem at bay is very strict policing of airports. Keep yourself aware, and Defra will issue bulletins when appropriate to all registered sheep keepers.

SCRAPIE

This is a notifiable disease. If you spot the symptoms of this condition, it is a legal requirement that you notify Defra immediately.

When affected by scrapie the sheep becomes nervous, jumpy, uncoordinated and ceases to be a flock member, preferring to stand on her own. Due to her lack of nervous coordination she will not be eating and will waste away. It is not common, but does occur. Some areas or even some farms seem to be prone to the problem.

If a sheep dies, even where you consider it to be a 'normal' death, there is still at the time of writing a scheme whereby, by phoning the appropriate organisation (ask Defra) between Monday and Thursday, your dead sheep may be collected free of charge and the brain examined for scrapie. The test result will nearly always be clear, and in effect monitors the scrapie situation in your flock.

MAEDI VISNA

This is a notifiable disease. If you spot the symptoms of this condition, it is a legal requirement that you notify Defra immediately.

Maedi visna is a relatively new disease, only hitting the UK in the last quarter of the 20th century. The sheep will stand alone, looking dull and listless, with copious nasal discharge and salivation. She will need to be culled.

WATERY MOUTH

This condition occurs in lambs, usually within the first few days of life. A hitherto healthy lamb will become dull, 'floppy' and cold. There will be little or no interest in suckling from its mother, and the lamb will probably be dribbling from its mouth. If you pick up the lamb and shake it gently the belly quite literally 'rattles', hence the alternative name of 'rattle belly'.

Watery mouth is caused by the lamb ingesting E. coli bacteria before it has consumed sufficient colostrum to give it immunity, so it is essential to ensure the lamb has this colostrum within the first hour of life. The main source of infection is dirty conditions when lambing, or a mum with a dirty fleece. Treatment with antibiotics may be successful if the condition is caught early.

JOINT ILL

The symptoms of this condition, which affects lambs, are hot and swollen joints. Eventually abscesses may develop. Treatment with penicillin can be effective if treated early. In more advanced cases it may be preferable to have the lamb put down.

The cause is a streptococcus organism that can be found, sometimes, on the skin of ewes. It can enter into the lamb via the umbilical cord, once again underlining the importance of clean bedding if lambing inside, and thorough dousing of the umbilical cord with iodine as soon as possible after birth.

ORF

This is a virus-induced condition, which results in scabs around the nose and mouth of affected lambs. It is very contagious and very serious if you handle the affected area and become contaminated yourself. If it occurs, the lamb will infect its mother, resulting in scabby, sore teats, mastitis and eventually a dead ewe. More ewes may become infected as the hungry lamb tries to 'cross-suckle'. You will need to consult your vet for advice.

Similar symptoms may be caused by grazing 'thistly' pasture and, although it appears the same, may not be infectious.

FEEDING

*Sheep will usually, but not always, find enough
to eat as supplied by Mother Nature. Sometimes,
however, they need a little help to make sure
they have the correct balance of nutrients
throughout the year.*

Always remember that a sheep was designed to eat forage,
be it grass, fodder crop (forage crop) or conserved grass
in the form of hay, haylage or silage. If you have a few sheep
on a relatively large area, Mother Nature will provide the grass
without the addition of bought-in fertilisers. The only time
that concentrates (bought-in, highly nutritious feed) may be
required is when your sheep are pregnant or lactating; to
encourage early lambs to grow quickly for the high-value
Easter market; or to add extra condition to a show sheep.
I will now go on to discuss the feeds that Mother Nature,
with a little help from you, can provide.

Grass is always there and readily available. You may find that a knapsack sprayer is useful on occasion to treat nettles, thistles or docks as they pop up their unwelcome heads. By using this method of control your pastures always look tidy, are weed-free, and your sheep can continue to graze the field.

Ragwort with its pretty, yellow, open, daisy-like flowers is the one weed you should pull up and burn. Sheep will not eat it, but if you make it into hay or silage, they may eat it unknowingly and it could kill them. Sheep do not often succumb to poisoning due to having eaten something they should not because they have a pretty good idea of what is an acceptable diet.

Most old pastures will have naturally occurring white clover, which is very beneficial. It provides nitrogen to stimulate grass growth, is highly nutritious and is a relatively good source of protein for your sheep. Red clover is a much bigger plant that does not occur to the extent of the smaller white clover. If there is no clover in the sward after very hard grazing in the spring, buy some clover seed and broadcast it by hand on the open turf. This is quite effective and is a cheap way to improve your grassland.

Old pastures will also have a selection of broad-leafed plants, usually deep rooted and a good source of mineral for the sheep; examples are plantain, meadow vetch, dandelion and chicory.

If you make hay from mature pasture it can smell wonderful, if well dried, baled dry and stored free of mould; when you open it in the middle of winter it is possible to smell summer! The best conserved feed is made when the grass is still leafy and the grasses have first started to flower; the digestibility of the feed will be close to 70. The digestibility index is an indicator of percentage (in this case 70) of the nutrients in the grass that are available to the animal after digestion. 70 is a good feed, on which sheep will make oceans of milk when feeding lambs. However, to make hay of grass like this requires skill, experience and perhaps a lot of luck.

If making your own hay, haylage or silage, it is imperative that the sheep be removed at least six weeks before cutting is anticipated, when the field is bare. However, if the grass is growing faster than your sheep can eat it, two weeks may be enough. Ask the contractor who is going to cut your grass for the best advice.

FORAGE CROPS

Forage crops for sheep can include kale (the thousand-head variety is preferable, as opposed to marrow-stem kale), rape or fodder turnips. These involve ploughing the ground and so, to be practical, they need to be grown in areas of at least a hectare. Today's contractors have such large machines that even a hectare is probably too small an area. This means that, for most people, forage crops are probably not an option.

CONCENTRATE FEED

Concentrate feed is given to sheep in small quantities. It can be sheep cobs that can be put straight onto the ground, nuts or coarse mix (all the same ingredients as a nut but not compressed into nut form – it usually smells wonderful). As mentioned earlier, pregnant ewes may need concentrates, as may lactating females.

Other animals that may need concentrates in varying amounts include: early lambs growing rapidly for the early market; pedigree lambs that need to grow fast and develop solid flesh quickly to look their best in the show-ring; ram lambs that need to grow quickly so they are big enough to perform when the breeding season arrives; and mature rams that need to be built up to be capable of servicing 50 or more females in a 16-day period in the autumn.

It is imperative that sheep always have access to roughage when eating concentrates because, to achieve the higher level of performance you are aiming for by feeding the highly concentrated forms of nutrition, you are deviating from the natural diet to which the animals' digestive system has adapted, which is grass.

In extreme circumstances, when all your best efforts have failed and you find yourself in the middle of winter, surrounded by hungry sheep on bare frosty fields and there is no hay left in your barn, your sheep will perform well on a diet of straw (wheat, barley or oat straw, but wheat is the cheapest) and an increased ration of concentrates. You may raise your eyebrows in surprise but all the dietary

requirements are there – fibre in the straw keeps
the rumen (first part of the stomach) working and all
the energy, protein and mineral requirements can be
supplied by the concentrates.

A YEAR OF FEEDING SHEEP

To perhaps put a degree of sequential logic into the situation,
I will now go through a year of feeding sheep.

In late spring and early summer ewes and lambs should be
grazing, normally with grass in abundance. Provide the sheep
with a mineral bucket: these are obtainable from any merchant
and contain all the minerals that are required. Even the young
lambs will help themselves. At this time the rams will be,
ideally, away from the flock. If isolation is a problem, they
could be kept inside with access to hay as required and
water, but this is not ideal.

In late summer, wean the lambs, preferably on to clean
pasture. This can be as soon as the lambs are 12 weeks old.
The amount of nutrition they obtain from mother's milk is
negligible by this time.

Three to four weeks before the ewes go to the ram, they
need to go onto a rich, lush pasture. As this fades, put hay
in the racks and this is all that is needed until the ewes are
six to eight weeks pre-lambing, when the concentrate may
be provided.

The ewes can roam over your entire pasture area in the winter
months as this is, believe it or not, beneficial to the grass.

Grass photosynthesizes as long as there is light, which produces sugars. At the same time, grass will only grow if the temperature is above 5.5°C (42°F). If grass is allowed to remain long during the autumn and winter, the net effect is for the cell sap to produce a very diluted sugar solution – namely lots of cells, but not much photosynthesis due to short day length. If the sheep have eaten the grass to the point of it being very short, the net effect is to concentrate the cell sap, which lowers the freezing point. This means that the cell sap is less likely to freeze, the cells will not burst, and the grass will not be killed by the frost; when spring comes the grass will grow at a phenomenal rate, looking like the picture-book pasture you always dreamed of.

The level of concentrates fed to the ewes will be about 100g per ewe about six weeks before lambing, raised to 750g per head about two weeks before lambing. If the ewes are lambing in mid-winter or early spring when there is no grass, maintain the level of concentrate and always keep the hay racks full. Once the grass is in abundance the concentrates can be reduced to zero, but maintain the mineral supply as mentioned earlier. If you have sufficient numbers, triplets can be left on the ewes and the ewes fed in a separate group, on grass and concentrate. Only leave triplets on the ewe if all three are doing well; any sign of weakness remove one lamb, leave the ewe with twins and foster or bottle rear the lamb.

LAMBING AND ASSOCIATED TASKS

Lambing time is the most exciting, heartbreaking, exhilarating, disheartening and demanding period in any shepherd's year.

My first involvement in lambing was at the age of five, when I was dispatched to suckle lambs that were bottle fed. At seven, I was to take the dog and make sure that the ewe stood still and allowed her fostered lambs to feed. At ten, I had to walk through the ewes at 7.30am while my dad was milking. I overslept on one occasion and a ewe had triplets, two live and one dead. Of course it was my fault – I did not oversleep again.

This season was my 47th lambing, and I still feel excited as the time approaches for the first lamb to appear. What causes this excitement? The wonder of new birth, still magical after all these years, particularly if you happen to be there to see a lamb born, totally and completely unaided. The ewe takes a deep breath, staggers to her feet, and begins to lick the lamb. How can a wet tongue, licking a wet lamb, cause it to become dry? Strange, but it does, far more effectively than any amount of rubbing with towels.

CHECK THAT THE EWES HAVE BEEN SERVED

It is useful to be able to tell when each ewe has been served. The best way to do this is to get the ram fitted with a harness around his shoulders to which is attached a marker crayon, on his breast bone between the fore legs. Ideally fit a pale colour, say yellow, for the first ten days, then perhaps blue for the next ten days, and finally red. If a ewe is served twice, blue can be seen over yellow, and red over blue, but not if used in the reverse order.

Twenty-one weeks is the gestation period, so a 5th November service results in a 1st April lambing.

PREPARATION FOR LAMBING

The degree of preparation depends on your system. If lambing is to take place in late April or May, a few 'bonding' pens, or isolation pens, will be needed. These will preferably be under cover and with electric light so you can inspect your problem ewes during the night without disturbing them with a flashing torch. There will rarely be a season when you do not need these pens, no matter how few ewes you have.

If you lamb between late December and March, there is a high probability that the weather will be wet and cold. There will be no dry ground for the newborn lambs to fall onto as they are born and no warm sunshine to dry them and warm them through. This means you will need sufficient space to house your ewes before they lamb (the official recommendation is $3\frac{1}{2}m^2$ (12ft²) minimum per ewe), then you need space to put them after they have lambed.

If indoor lambing, each ewe ideally needs a minimum of 24 hours to become totally acquainted with her lambs and them with her. If left in the group, there is a high likelihood that you will walk into the lambing yards to find three ewes lambed; one with no lambs, one with one and one with five lambs, when in fact they have had two each. In a field there is enough space for a ewe to wander off to 'her spot' where she will 'paw' and sniff the ground, lie down, stand up, look incredibly uneasy and eventually pop out her lambs – away from the others. She may stay there for 24 hours before rejoining the flock, but the bond has been formed. Mum knows the children, and the children know mum. As someone once said, 'How many bleats to the baa.'

Your first task upon finding newborn lambs (preferable if born outside, but essential if born indoors) is to thoroughly douse the lambs' umbilical chord with iodine. This will minimise the risk of infection entering the lamb via the umbilical cord before Mother Nature has caused it to heal naturally. In preparation for this task buy a small (at least 1-litre but preferably 5-litre) container of concentrated iodine solution and dilute it three-to-one into an old washing-up liquid bottle.

Always, after you have finished lambing and the weather enables you to leave the sheep outside all the time, clean, hose out and disinfect your sheds, ready for next year.

A good idea, if you have never lambed sheep before, might be to lamb your ewes a short while after a local farmer has lambed his ewes. Ask if you can help. This does not mean cuddling the poorly lambs – if a lamb is ill, you should treat it with whatever is necessary, make sure it is fed and warm, then leave it until the next feed is due, casting an eye in its direction if you have the opportunity. Your energy and efforts must be primarily reserved for the healthy lambs.

THE MAIN EVENT

When a lamb is born naturally, you will see two front toes first, closely followed by a nose. It is very similar to a diver, diving into a swimming pool, head tucked in behind two arms (see diagram, below). If the ewe needs help, the feet can be very slippery to grip – fine lambing ropes, slipped over the leg and behind the first joint, can be very useful in such cases. Always pull on both legs and, if born forwards, the head and two front legs must all come together.

The nose and one of the two feet of a normally presented lamb.

A lamb halfway out through the effort of the mother's contractions.

The mother starts to lick her newborn lamb.

The lamb raises its head in response to the mother's licking.

If there is no head evident, the lamb may be positioned backwards. How do you tell? If forwards, the two little digits behind the front feet (look at an adult sheep to see for yourself) will be underneath and, if backwards, they will be on top. You will need to feel for this inside the sheep.

If you think the lamb is backwards, follow the legs back and find the hock (joint in the lower part of the hind leg). Make sure both back legs belong to the same lamb, check for the tail and, having done all that, gently pull out the lamb.

If the lamb is positioned forwards and the head is back, or if the head is presented with no front feet, the lamb has to be pushed back, firmly but gently, into the womb. Once inside the pelvic ring you have to find the head by feeling along the legs to the shoulder and pull it round; the same for the legs if you have only a head. You then, if the ewe is not helping much with natural contractions, which can be very strong, gently ease the lamb out by gripping the lower jaw between your thumb and index finger, one leg between index and middle finger, and the second leg between middle finger and fourth finger. It is often said that the only thing that beats a successful difficult lambing is seeing your own children born.

APPARENTLY LIFELESS LAMBS

On occasion, lambs may be born that appear lifeless. To stimulate them to take that first breath, just as newborn babies may be smacked on the bottom, pinch an ear between the fingernails of thumb and fingers, prod the end of nostril with a straw or massage the chest by rotating the shoulder joint. Rub the lamb's chest and head with some clean, dry straw.

The final aid to stimulating life is in the case of backwards born lambs. They may have fluid on their lungs. To be sure that they have not, grip them by their hind legs and swing them in a circular motion, thus forcing the fluid out of the lungs. Retain a firm grip – letting go of the lamb at this point will do little for its life expectancy.

AFTER LAMBING

WORM DRENCHING AND FEET TRIMMING

Once your sheep have lambed, it is the ideal time to treat them with one of their two annual 'worm drenches'. I also use this time as an opportunity to trim their feet. It is virtually the same as cutting our toenails. Again, a little practice on a friend's or neighbour's flock is a good idea. Sit the sheep on its rear end and lean it back against you. By this means you can grip the sheep's back between your knees and at the same time lean forward to grip the sheep's foot in one hand. Hold the foot-trimming pliers in the other hand and trim the feet (see diagram).

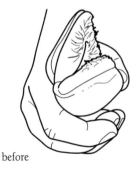
before

after

Commercially available foot-trimming crates can be purchased, which rotate the ewe into a position where the feet are readily accessible, but such a purchase may be difficult to justify for a small flock. Your back, however, may beg to differ.

TAGGING

All sheep must be tagged and if they are to be used for breeding and expected to live beyond 12 months, they must be double tagged with one tag in each ear. If they are destined for the table, however, then one tag will suffice as the law stands at the moment. Tagging is a legal requirement and easiest to carry out soon after birth.

I double tag all our sheep, because a 15-month-old wether will supply you with lamb chops that are more like T-bone steaks and are unbeatable. Add a little homemade crab apple jelly, roast potatoes, garden peas and a glass of homemade elderflower champagne; you will think you are in heaven.

DOCKING AND CASTRATING

Your lambs' tails must also be docked soon after birth. The rubber ring is placed about three finger widths from the base of the tail – enough to cover the anus and vulva. This distance is a legal requirement.

Some breed societies, such as the primitive breeds Soay, Shetland, Manx Loghtan and Hebridean, stipulate that the tail must not be docked, and by tradition White-faced Woodland males have tails, while females do not.

Ram lambs must at this time also be castrated if not required for breeding. This avoids a whole mass of problems in the autumn when age, maturity and hormones take over. Some people find this difficult; a training course or practice with a friend's or neighbour's flock, under their supervision, can be a good idea.

TOOLS

In conclusion, there are some essential tools that you will need during lambing time, and it is a good idea to have everything prepared ahead of the main event. The tools you will need are:

- Drenching gun for the worm drench, which is usually supplied with the product.
- Applicator for pour-on fly repellent. Again, this is usually supplied with the product.
- Foot-trimming pliers.
- Antiseptic aerosol or spray for feet.
- Iodine.
- Lambing ropes.
- Plastic elbow-length gloves.
- Ear tagging pliers and tags.
- Castration pliers and rubber rings.

FOSTERING
TECHNIQUES

*A healthy ewe should have two functional teats,
which is the best method of feeding young lambs.
Fostering makes full use of this readily
available resource.*

You might imagine that information about fostering
should appear in a chapter devoted to lambing
but I believe it is so important that it merits a chapter
of its own.

If you are lambing sheep, it is a fact that you will eventually
find yourself in a situation where a lamb cannot be fed by
its own mother. This could be for any one of a number of
reasons: the ewe may have died after a difficult lambing, the
ewe had no milk, she had two lambs and only one functional
teat, she had three or four lambs. It is amazing how many
combinations can result in the same conclusion: the lamb
needs milk.

It is pertinent at this point to emphasise that any sheep
that gives you problems regarding milk supply should
be clearly marked. She should be sold as barren once
she has put on some weight. If you are 'kind' and decide
to keep such a sheep, and we have all done it at some
time, you are only storing up problems for next year.
If your sheep is only there as a 'green' lawnmower, OK,
but don't put it in lamb again – it will only bring more
problems and heartache.

First and foremost, the lamb must have milk as soon as possible after birth, ideally within six hours of birth. The first milk the ewe produces is known as colostrum. It is very thick, creamy and yellow in colour. It carries the complex protein molecules, which are the antibodies that will give the lamb resistance to the challenges it will face from the environment. If the lamb passes faeces that are relatively firm and a dark yellow in colour, you know that it has had colostrum.

Ideally, a lamb should have its mother's own milk but, if not available, milk from any other newly lambed ewe will suffice. It is a very good practice to take colostrum from the first ewe that lambs a single. You will be able to identify which teat the single lamb is suckling from, as the udder will be soft on that side. Take the milk from the full side and don't be afraid to empty it as the ewe will soon make some more. Taking milk from the full side is quite important. First of all it is easier, but secondly a lamb will sometimes only suckle from one side. If you have taken milk from the one side it feeds from, you are doing the lamb no favours.

Store the colostrum in the freezer. Small margarine tubs are useful as containers. When you need it, heat it up slowly. Overheating causes it to coagulate; this will not only block up the teat on your bottle but also denature the proteins, causing them to be ineffective as antibodies.

Keep your eye open for a ewe with a large udder bursting with milk, even if she hasn't lambed. You will find her very useful if milk is needed urgently for a new-born lamb.

It is important to feed the lamb as much colostrum as it will drink in the first hours of life. The absorption abilities of the stomach wall decline very quickly in the first 24 hours of life, rendering the benefits of colostrum virtually ineffective 24 hours after birth.

FINDING A FOSTER MOTHER

Having managed to find a source of colostrum, the next stage is to find a mother for your hungry, motherless lamb. Of course it is quite feasible to rear the lamb on bought-in milk substitute, but the powdered milk is not cheap! Anyway, if a ewe is there with the milk, you may as well use it.

If a ewe has lambed with stillborn lambs it is quite easy to turn her into a foster mother. Having just lambed – minutes ago is ideal – the mothering instinct is still strong. If you rub your motherless lamb with the wet, dead lamb, the motherless lamb will itself become wet with the fluids of the foster mum's own lamb or lambs. Take the dead lamb away and pen the ewe up in one of your already prepared pens with the orphaned lamb – she will hopefully lick it, thinking it is her own.

If a ewe loses lambs after two or three days, it is more difficult to get her to foster. A ewe with no lambs at this stage will eventually accept a foster lamb, but this may necessitate holding the lamb to the ewe on a regular basis every three or four hours, for as long as it takes the lamb to fill up with ewe milk. You may need to be quite firm as the ewe could well object in a considerably violent manner.

FOSTERING ON A EWE WITH
A SINGLE LAMB

The most difficult fostering, and perhaps the most rewarding, is to foster a lamb onto a ewe that has its own single, healthy lamb. If your sheep have been scanned you will know which ewe will have singles and it is simply a matter of being aware and on-site when she is giving birth. If your ewes are not scanned and you have lambs to foster, be there at each and every lambing to check if there are any more. One indicator of a single is a huge lamb, compared with the twins already born to other ewes. You can also examine the ewe internally.

Once you have discovered there is no second lamb, hopefully before the ewe has stood up after the first lamb, have a helper close by with the foster lamb. Then, making a fist of your hand, insert it into the birth canal and let her internal muscular contractions force your fist out. She will think she has had a second lamb. Meanwhile your assistant is soaking the foster lamb in fluids from the ewe's newborn lamb. By placing the foster lamb on the ground behind the ewe as your fist is being ejected, there will be more fluids, and hopefully some placenta on the foster lamb. Finally, firmly but gently, tie the diagonally opposite legs on the foster lamb together with string, for about 45 minutes. Allow the ewe to rise and she will, in a very energetic manner, lick her twins – you hope.

This technique has rarely failed with our sheep; the string helps because if the newborn (foster) lamb immediately jumps up to its feet and proceeds to run around the pen like an Olympic sprinter the ewe may perceive that something is not quite as it should be.

Keep close attention for some time and, assuming all goes well, remove the strings from the foster lamb. If the ewe is not particularly keen on her second lamb we have found that tying the ewe in a corner of the pen on a relatively short halter can be effective in helping the fostering process. She can reach her feed and water, she can stand up and lie down, but she is less able to butt the lamb of which she does not approve. She eventually accepts the fact that she has two lambs pulling her teats instead of one and, provided you ensure that the lambs are actively feeding, it releases you from the onerous and time-consuming task of standing with the reluctant ewe numerous times a day. And when you see the ewe walking happily across the field with two full, healthy lambs skipping behind her, you will realise that all your hard work was worthwhile.

POST-LAMBING
MANAGEMENT

*The time is approaching when you can lean
on the gate and view the results of your months
of hard work and planning.*

Once the ewe and her lambs have accepted the fact that they are a family and they have joined the group, be it in the loose yard or in the field, your level of attention has to adjust accordingly. How do you know which lamb belongs to which ewe? Before releasing the ewe and lambs into the group, number them with the same number on the ewe and her lambs, plus a number of dots that relates to the number of lambs. For example, a ewe with twins could be numbered 23. Her lambs are numbered 23, and she and her lambs have two dots on, so you always know she should have two lambs with her.

HANDLING EWES

Cast your eye over your animals whenever you walk or drive past, checking for anything that does not look 'right': a ewe on her own; lambs not joining in play with the others. If a ewe has three lambs, ensure all three are thriving. Even twins or singles may not always thrive. If a lamb looks hungry, catch the ewe to discover why.

Your ewe may be old and not making enough milk; she may be young but lazy and not a natural mother; she may have an udder infection, in which case she will need treating. This will usually require a combination of stripping out the infected milk and treating the ewe with antibiotics. This is smelly, unpleasant for you and the ewe, and difficult in that the ewe's udder will be sore due to the infection and associated inflammation – she will not appreciate the attention, even though it is for her benefit.

For any task involving the handling of a ewe, a sheep halter can be useful in that at least one end can be kept relatively under control. If handling a ewe in a pen, back her into a corner, place your foot on the floor in front of her near shoulder and place your knee in her shoulder at a slight angle, effectively wedging her into the corner, with your head pushed into her flank, and your other knee on the floor. She is fixed, you are stable, and both hands are free to handle lambs that are unwilling to suckle, or to strip milk from the udder.

FEEDING AND WEANING

Make sure your flock have an abundant supply of forage, be it grass, hay, haylage or silage. As the season progresses, and grass is in abundance, they will not need hay or concentrate – unless feeding threes (see page 67).

Lambs grow at an amazing rate. They can be weaned at 12 weeks, but can be left longer. When you do eventually wean the lambs it may be beneficial, but is not absolutely necessary, to worm them with anthelmintic. Leave them inside, on straw to keep them clean, with water to drink and perhaps a little hay in a rack while the anthelmintic takes effect. After 24 hours release them onto clean pasture.

KILLING LAMBS

Once a Down lamb or a commercial lamb has reached about 36kg (80lb) live weight, it is ready to be sold for slaughter. The huge Longwools – Wensleydale, Leicester Longwool and Lincoln Longwool – may be heavier before they are ready for slaughter.

If you have a small flock, surplus (castrate) males may be sold to friends for the freezer. You will have to find a local butcher to help you with slaughtering and butchering them. This is quite a common occurrence, and butchers are familiar with what is required. (For more on selecting lambs for slaughter, see page 84.)

If you do not sell to friends you may choose to take your lambs for slaughter to a local market. If you have never sold 'killing lambs' before, it may be advisable to ask someone with experience to help you. Remember that the individual you ask could well have a lifetime of experience; be prepared to offer something in return for this help; think what a consultant might charge you for advice on your business. An experienced shepherd or farmer is no less experienced than your consultant, probably more so, so treat him accordingly.

WHEN IS YOUR LAMB FIT FOR SLAUGHTER?

It is all in the fingertips. As I have already said, the ideal weight will vary. The commercial sires, and the larger Down breeds (except South Downs, which are smaller) will be best slaughtered between 16–20kg (35–44lb) dead weight, which equates to 35–45kg (77–99lb) live weight. Up to 30–66kg (66–145lb) live weight is acceptable.

If you are taking your lambs to a local market, the buyers are looking for a well-fleshed lamb, weighing 18kg (40lb) on the hook – about 40kg (88lb) live weight when it left you. If it is for your own freezer, the choice is yours. My personal preference is for higher weights because, obviously, the bigger the lamb, the bigger the joints.

To check their readiness, put your hand across their back, side to side, and gently move your fingers from front to back. If what you feel is similar to the back of your hand, the sheep is too fat. With your fist clenched, if the back feels like the first joint of your fingers, the sheep is perfect, but if it feels like your knuckles, it is too thin. The primitive breeds will be much lighter, while the Oxford, Longwools and Wensleydales will be heavier when ready for slaughter. Always remember, the more mature the lamb at slaughter, the better the flavour. You just need to cook it in the oven perhaps a little longer.

Those lambs not sold for slaughter but retained for breeding can complicate your grassland management. Your ewes, once their lambs are removed, will only need a relatively bare pasture as they dry off, and little more, as long as they remain in good body condition, until about three weeks before they join the ram. Any ewes with only one effective teat or any older ewes should be culled at this point. Any ewes that are a little lean, due to perhaps having fed triplets, or fed twins particularly well, may need more lush pasture. Once the critical three weeks pre-tupping has arrived, put the ewes onto lush pasture, with the ever present mineral bucket, so that they are gaining weight and rising in condition. This will ensure optimum ovulation and a good consequent crop of lambs next year

Meanwhile, the killing lambs and female flock replacements are growing well on their lush pasture. If you have retained ram lambs for breeding, they need to be segregated at weaning and offered concentrate feed so that they are large, powerful and aggressive when the breeding season arrives. They should not be in the same group as the mature rams who will bully them at the feed trough, scoff their ration and therefore prevent the ram lambs from growing as they should.

SHEARING

*It's a sad reflection on our society's
requirement for convenience that wool
for clothing is not as popular as it once was,
with the result that it will cost you far
more to shear the sheep than you will
receive for the wool. There can be few things
more satisfying, however, than wearing
a jumper made from your
own sheep's wool.*

There are, believe it or not, two breeds of sheep that do not need to be sheared: the Wiltshire Horn and the Easy Care sheep, which is based upon the Wiltshire Horn. Otherwise your sheep need to be sheared as soon as the weather becomes warmer in late April or May. If sheared too soon and the weather becomes colder, a lactating ewe will respond by producing less milk; leave it too late and fly strike will become a problem. Of course, as soon as conditions are perfect for you they are also perfect for everyone else, so finding a contract sheep shearer may not be easy.

Once you have persuaded your contactor that his life will not be complete until he has sheared your sheep, arrange your handling system and staff in such a way that he will want to come back again. Consider what you are charged for a haircut; the sheep could weigh anything from 40kg (80lb) – small and awkward – to 120kg (265lb) – big and awkward. Shearers have to catch the ewe, turn it over, hold it while he shears it and do a good job – all for around £1.50. He has to work hard for his money, and if the sheep are ready and waiting when he arrives, with a handling system and convenient electrical power point that are to his liking, when you phone next year, you will have more chance of him coming to shear your sheep when you want them to be sheared.

A good shearer can work extremely quickly – an unofficial record of around 760 sheared in nine hours was set in 2008.

USEFUL ADDRESSES

Defra (Department for Environment, Food & Rural Affairs)
Nobel House
17 Smith Square
London, SW1P 3JR
Tel: 08459 335577
Email: helpline@defra.gsi.
  gov.uk
(To find your local Defra office contact them at their headquarters, given above.)

Local Trading Sandards Office
1 Sylvan Court
Sylvan Way
Southfields Business Park
Basildon
Essex, SS15 6TH
Tel: 0845 4040506
www.tradingstandards.gov.uk
(To find your local trading standards office contact them at their headquarters, given above.)

National Sheep Association
The Sheep Centre
Malvern
Worcestershire
United Kingdom
WR13 6PH
Tel: 01684 892661
www.nationalsheep.org.uk

British Wool Marketing Board
Wool House
Roydsdale Way
Euroway Trading Estate
Bradford
West Yorkshire, BD4 6SE
Tel: 01274 688666
Email: mail@britishwool.
  org.uk

BRITISH SHEEP SOCIETIES AND ASSOCIATIONS

The below is a selection of societies and associations for individual sheep breeds. For further contact details, see the National Sheep Association website (see page 88).

Beltex Sheep Society
Shepherds View
Barras
Kirkby Stephe
Cumbria, CA17 4ES
Tel: 017683 41124
Email: info@beltex.co.uk

The Society of Border Leicester Sheep Breeders
Rock Midstead
Alnwick
Northumberland, NE66 2TH
Tel: 01665 579326
Email: info@borderleicesters.
co.uk

British Bleu Du Maine Society
Long Wood Farm
Trostrey, Usk
Monmouthshire NP15 1LA
Tel: 01291 673816
Email:jane@bleudumaine.
co.uk

The Blue-faced Leicester Sheep Breeders Association
Riverside View
Warwick Road
Carlisle, CA1 2BS
Tel: 01228 598022
Email: info@blueleicester.
co.uk

The British Charollais Sheep Society
Youngmans Road
Wymondham
Norfolk, NR18 0RR
Tel: 01953 603335
Email: office@charollais
sheep.com

**Dorset Horn and Polled
Dorset Sheep Breeders'
Association**
Agriculture House
Acland Road
Dorchester
Dorset, DT1 1EF
Tel: 01305 262126
Email: mail@dorsetsheep.org

**British Friesland Sheep
Society**
Secretary: Mrs L Baber
Weir Park Farm
Waterwell Lane
Christow
Exeter
Devon, EX6 7PB
Tel: 01647 252549

Hebridean Sheep Society
Secretary: Mrs Helen Brewis
Coney Grey
Gun Lane
Sherington
Newport Pagnell, MK16 9PE
Tel: 01908 611092
Email: info@hebridean
   sheep.org.uk

Lleyn Sheep Society
Society Secretary: Gwenda
Roberts
Gwyndy
Bryncroes
Sarn
Pwllheli
Gwynedd, LL53 8ET
Tel: 01758 730366
Email: office@lleynsheep.com

**Oxford Down Sheep
Breeders Association**
Secretary: Paul Froehlich
Hillfields Lodge
Lighthorne
Warwickshire, CV35 0BQ
Tel: 01926 650098
Email: secretary@oxford
   downsheep.org.uk

**The Southdown Sheep
Society**
Meens Farm
Capps Lane
All Saints
Halesworth
Suffolk, IP19 0PD
Tel: 01986 782251
Email: secretary@southdown
   sheepsociety.co.uk

Suffolk Sheep Society
Unit B
Ballymena Livestock Market
1 Woodside Park
Ballymena
Co. Antrim
Northern Ireland, BT42 4HG
Tel: 0282563 2342
Email: lewismcc@suffolk
   sheep.org

**Swaledale Sheep Breeders
Association**
Barnley View
Town Head
Eggleston, Barnard Castle
Co. Durham, DL12 0DE
Tel: 01833 650516
Email: jstephenson@
   swaledale-sheep.com

British Texel Sheep Society
National Agricultural Centre
Stoneleigh Park
Kenilworth
Warwickshire, CV8 2LG
Tel: 024 7669 6629
Email: office@texel.co.uk

**Wensleydale Longwool
Sheep Breeders Association**
Secretary: Dr D.L. Clouder
   (Lynn)
Coffin Walk, Sheep Dip Lane
Princethorpe
Rugby
Warwickshire, CV23 9SP
Tel: 01926 633439

FURTHER READING

British Sheep, ninth edition
(The National Sheep
Association, 1998)

The Showman Shepherd, David
Turner (Farming Press Books
and Videos, 1990)

*British Sheep Breeds: Their
Wool and its Uses* (The British
Wool Marketing Board,
1978)

GLOSSARY

Barking A sheep (or rabbit or hare) stripping the bark off a tree is said to have 'barked' the tree. If the bark is removed around the entire circumference of the tree it will die.

Commercial sheep A generic term that applies to the plethora of breeds and crossbreeds used by everyday farmers in the production of lamb for the table.

Conformation The shape and style of a sheep. In a hill ewe, the type will be smaller and probably finer than a heavily muscled sire of lambs for the butcher.

Draft ewe A ewe that is sold out of the flock because she is no longer required.

Dagging Removal of wet faeces, soaked into the fleece at the rear of the sheep. It may be due to wet and lush spring or autumn grass, or it may indicate that the sheep has scour due to a worm infestation in the gut.

Docking The term refers to putting a rubber elastrator ring on a lamb's tail. It is a legal requirement that the 'dock', the tail that remains, must cover the anus and vulva. If it is slightly short judicious trimming of the tail wool can overcome the shortage.

Down sheep Traditional heavy breeds of sheep, native to England's 'shire' counties. Includes Oxford, Shropshire, Dorset, Hampshire, South and Ryeland.

Drenching Administering a medicine down an animal's throat.

Forage crop Traditionally, kale, oil seed rape or stubble turnips, sown after the harvest of cereals, usually in the Eastern Counties or the South of England, for the fattening of store lambs bought from hill farmers at the autumn sales.

Fly strike Flies lay eggs in hot, sweaty wool on the backs of sheep with thick, heavy fleeces, particularly the Down breeds, in hot and humid weather, or in the faeces-soiled wool at the rear end. The eggs hatch into maggots and eat the sheep alive.

Gimmer A sheep with two broad teeth.

Hard core A term used to describe material used to provide a firm surface; can be limestone or, more cheaply, recycled crushed/double crushed concrete, used for roadways, farm tracks or a hard surface on which to handle your sheep.

Hay Grass, cut when quite long, dried by the sun and wind and put into bales, after being turned for two to four days to assist in uniform drying of the crop. For small flocks hopefully a contractor can be found to bale small 30 to 40kg (66 to 88lb) bales.

Haylage Grass, semi-dried, baled and wrapped in an air-tight plastic wrap. Usually only turned once or twice.

Hefting The facility of hill and moorland flocks, pure bred on the same land for generations, to stay on their own piece of hill or moorland without fencing to retain them.

Hogg A sheep that is one year old; it may be female or a castrate male, hence the terms 'hogg' and 'ewe hogg'.

Hogget See Hogg (above).

Killing lamb A term used to describe lambs that are destined for slaughter.

Killing-out percentage The proportion of saleable meat on a lamb once it reaches the butcher.

Lambing percentage The number of lambs born per 100 ewes. In the hills, a lambing percentage of around 100 per cent is acceptable (100 lambs born to 100 ewes). On the milder lowlands however, around 200 per cent would be ideal (200 lambs born to 100 ewes).

Mutton The term used to describe meat from a mature sheep.

Pastern The rear or 'heel' part of a sheep foot.

Pleaching To cut a vertical growth of hawthorn or other hedge plant in such as way that it can be bent over nearly horizontal, which still attached to the base. A solid, stockproof barrier can be created but needs a skilled craftsman to do the job.

Poached Pasture land becoming damaged by the feet of livestock during wet weather. The damage can be considerable if done by cattle but if sheep are the cause, once removed to other pasture, the land will recover as winter turns to spring and grass begins to grow.

Poll The top of the head.

Primitive sheep Type of sheep supposedly left by the Vikings, such as Hebridean (St Kilda), Manx, Shetland and Soay.

Race A narrow, one-sheep's-width passageway through which sheep pass. Gates at the front and the back make it possible to contain sheep for vaccination and worm drenching tasks.

Ram The male of the ovine species.

Shearling A sheep that has been sheared once.

Silage Grass dried by the sun and wind; usually not turned. Large commercial farms use huge machines to store hundreds of tonnes of silage. On a small scale it can be baled and wrapped in the same way as haylage.

Store lambs Lambs that are intended for slaughter, but not yet sufficiently well fleshed.

Stripping out Removing milk or infected milk from a sheep's udder by squeezing it out of the teat.

Teaser A vasectomised male ram, used to induce ewes who have had no sight, sound or smell of a ram for the preceding 30 days to come on heat at the same time, this making the lambing period, in theory, much shorter.

Tegg Term used to describe a sheep with two broad teeth.

Terminal sire Terminal sires are rams from large, meaty breeds of sheep, which will produce a meaty lamb no matter how slim the ewe it is crossed with.

Tup Term used for the male of the ovine species.

Two-tooth, four-tooth, six-tooth, full-mouthed An indicator of age. A sheep has no teeth at the top of the front jaw, but eight at the bottom. Up to about 12 months, the milk teeth remain, but then, starting at the middle, the broad teeth emerge in pairs; hence a two-tooth is about one year old; a four-tooth is 18 months old.

Wether A castrate male lamb. This term is usually applied when the animal looks more like an adult sheep than a lamb.

INDEX

DEDICATION

I dedicate this book to Ken Pearson MBE for
services to agriculture; one of so many who gave so
much (1939–1945). To my tolerant and long-suffering
wife Rosemary, for putting up with her eccentric husband.
To my son Simon and daughter Katrina, the two people
who make everything worthwhile.

ACKNOWLEDGEMENTS

Amazingly the information in the preceding book has come
from the little grey cells. I must thank the academic staff of
Reading University for making the subject of sheep so very
interesting; our vets John Arnott (The Surgeon) and Hywell
Parry (Doctor Death) for answering my many queries over the
years; Kirsty for deciphering my scribbles and making it
legible to the publishers; Phil for computer help and advice
and photographs; Robert, for help and laughter shared; Colin
for use of computer and finally B.J. for a lifetime of friendship.